Assessing Revolutionary and Insurgent Strategies

UNCONVENTIONAL WARFARE STUDY
RESEARCH AND WRITING GUIDE

Paul J. Tompkins Jr., USASOC Project Lead
Joe Tonon, Editor
Erin Hahn, Joe Tonon, and Guillermo Pinczuk, Authors

United States Army Special Operations Command
and
The Johns Hopkins University Applied Physics Laboratory
National Security Analysis Department

ASSESSING REVOLUTIONARY AND INSURGENT STRATEGIES

The Assessing Revolutionary and Insurgent Strategies (ARIS) series consists of a set of case studies and research conducted for the US Army Special Operations Command by the National Security Analysis Department of the Johns Hopkins University Applied Physics Laboratory.

The purpose of the ARIS series is to produce a collection of academically rigorous yet operationally relevant research materials to develop and illustrate a common understanding of insurgency and revolution. This research, intended to form a bedrock body of knowledge for members of the Special Forces, will allow users to distill vast amounts of material from a wide array of campaigns and extract relevant lessons, thereby enabling the development of future doctrine, professional education, and training.

From its inception, ARIS has been focused on exploring historical and current revolutions and insurgencies for the purpose of identifying emerging trends in operational designs and patterns. ARIS encompasses research and studies on the general characteristics of revolutionary movements and insurgencies and examines unique adaptations by specific organizations or groups to overcome various environmental and contextual challenges.

The ARIS series follows in the tradition of research conducted by the Special Operations Research Office (SORO) of American University in the 1950s and 1960s, by adding new research to that body of work and in several instances releasing updated editions of original SORO studies.

VOLUMES IN THE ARIS SERIES

Casebook on Insurgency and Revolutionary Warfare, Volume I: 1927–1962 (Rev. Ed.)
Casebook on Insurgency and Revolutionary Warfare, Volume II: 1962–2009
Case Studies in Insurgency and Revolutionary Warfare—Colombia (1964–2009)
Case Studies in Insurgency and Revolutionary Warfare: Cuba 1953–1959 (pub. 1963)
Case Study in Guerrilla War: Greece During World War II (pub. 1961)
Case Studies in Insurgency and Revolutionary Warfare: Guatemala 1944–1954 (pub. 1964)
Case Studies in Insurgency and Revolutionary Warfare—Palestine Series
Case Studies in Insurgency and Revolutionary Warfare—Sri Lanka (1976–2009)
Unconventional Warfare Case Study: The Relationship between Iran and Lebanese Hizbollah
Human Factors Considerations of Undergrounds in Insurgencies (2nd Ed.)
Irregular Warfare Annotated Bibliography
Legal Implications of the Status of Persons in Resistance
Narratives and Competing Messages
Special Topics in Irregular Warfare: Understanding Resistance
Threshold of Violence
Undergrounds in Insurgent, Revolutionary, and Resistance Warfare (2nd Ed.)

SORO STUDIES

Case Studies in Insurgency and Revolutionary Warfare: Vietnam 1941–1954 (pub. 1964)

TABLE OF CONTENTS

INTRODUCTORY MATERIAL

Introduction

The purpose of this writing guide is to assist individuals tasked with writing case studies that examine unconventional warfare (UW). It is a companion to the *Insurgency Study Research and Writing Guide*, and both were developed by the National Security Analysis Department (NSAD) of the Johns Hopkins University Applied Physics Laboratory (JHU/APL) under the direction of the US Army Special Operations Command (USASOC), G-3X Special Programs Division. Over the past several years, JHU/APL has written several Tier I Insurgency Case Studies for the Assessing Revolutionary and Insurgent Strategies (ARIS) program to provide members of US Special Forces with an expanded body of knowledge on insurgency that is operationally relevant and supported by rigorous academic research. This body of work is now being expanded to include UW case studies. Both the research and writing guides were developed so first-time authors of case studies, whether analysts at JHU/APL or those within US Special Forces, have helpful instruction on the thought and approach used to develop insurgency and UW cases and to ensure uniformity among studies within these two categories.

The focus of UW case studies, and therefore this guide, is on the activities of the external supporter that enable resistance movements. The guide uses the term *resistance movements* broadly in order to include groups in the very early stages of development (e.g., engaging in nonviolent practices) up through established insurgencies. The term is used with the understanding that the author will clearly characterize the resistance movement central to the chosen case study and will detail how that characterization may have changed over time but focus the study on the motivations and methods of the external supporter of the resistance.. In this way, this guide differs from the *Insurgency Study Research and Writing Guide*, which focuses only on insurgencies and not groups in other phases of development and emphasizes the insurgent group itself and not the actions of the external supporter.

1

Providing external support to resistance movements has been a common method used by states to achieve various strategic political objectives, such as destabilizing rivals, increasing regional influence, and bringing about regime change. From the US perspective, external support in the form of UW is defined as "activities conducted to enable a resistance movement or insurgency to coerce, disrupt, or overthrow a government or occupying power by operating through or with an underground, auxiliary, and guerilla force in a denied area."[1] Other countries providing external support do not use the term *unconventional warfare* or assign a similar definition. Therefore, a given UW case study will look at external support from the perspective of the state providing it (e.g., Iran supporting Hizbollah) and seek to discern important lessons that can be drawn from studying the particular case. The US definition of undconventional warfare mentions an underground, an auxiliary, and a guerrilla force, but an outside actor may not use one of these forces or any combination of those components to support a resistance movement. Nonetheless, both the successful and unsuccessful methods of support used by other actors can be studied and applied to increase UW knowledge and best practices within the Special Forces community. It is worth noting that outside support can be provided by nonstate actors (e.g., diasporas or refugees). However, states remain the overwhelming providers of external support and are the focus of examples provided in this guide.

To understand both the nature of the external support and the provider of that support, it is first important to understand the strategic context in which the resistance movement emerged and the reasons for the external supporter to provide backing. Therefore, the guide recommends dividing the case study into six sections. Section I, the Introduction, should provide readers with a brief glimpse of the overall report, to include a description of the research questions that will be analyzed and the methodology used to answer these questions. Section II (The External Supporter) should provide an overview of the external supporter to introduce the reader to what will be the central focus of the study. Section III (The Historical Context) should address the historical and socioeconomic factors that shaped the environment in which the insurgency took place, such as the physical environment, socioeconomic conditions, vulnerabilities in the

2

population, and the extent of government control. These variables are particularly relevant to understanding the strategies and vulnerabilities of the resistance movement and the external supporter's assessment of those factors in its own decision making. Section IV (The Resistance Movement) should include a discussion of the variables that describe the resistance movement, particularly its structure and activities. Section V (External Support) should focus on describing the nature of the external support in detail and its effectiveness. Last, Section VI, the Conclusion, should be the concluding section and, in particular, should offer observations based on the information and findings provided in the case study. This guide is intended to provide guidance to authors in writing Sections II through VI.

Taken together, these sections will describe how and why an external supporter selected the resistance movement as well as the ways in which the external supporter enabled and leveraged the resistance movement. The writer should ensure that each section's content is relevant to the examination of the activities of external support.

The variables described within each section of this guide are not exhaustive but rather are factors commonly associated with the development of a resistance movement and the key considerations of an outside actor contemplating how to support a movement. They are designed to help the author think through the case study design and content and point to major themes to consider in the case study's development. In the course of research and drafting, it is important for the author to retain focus on how the information being provided is relevant to explaining external support and be mindful to emphasize this throughout each section. It is recommended that at the end of each major section, the author include a summary of how the content is relevant to understanding the external supporter. The author should consider the following three questions:

1. What are the vulnerabilities of the resistance movement and the larger population of which it is a part?

2. What is the extent of government control in the area and over the population, and what elements help measure or explain the level of control?

3. What is the level of control of the resistance movement in an area and within its leadership?

These questions will help the author focus on the vulnerabilities an external supporter can exploit or the difficulties it may face in providing support. In each section, the author should explain how strengths or weaknesses of certain variables help quantify the effectiveness of the support, which can feed the analysis in the concluding section.

The guide is written in a manner accessible to a broad audience, including those with research backgrounds and those undertaking their first major case study. It is meant to be practical and provide information that allows an individual to write a case study from beginning to end. It is not meant to illuminate major theoretical concepts or ideas. These concepts will be confronted by the author during the research process and will be unique to each case study. The extent to which they should be discussed is at the discretion of the author.

Purpose

The purpose of this guide is to help standardize the production of UW case studies under the ARIS research program. To this end, it enumerates a number of key variables that are central to the development of understanding outside support to a resistance movement seeking to disrupt, coerce, or overthrow a sitting government or occupying power. In particular, the purpose of this guide is to provide guidance to authors in researching and writing Sections II through VI of a UW case study.

At the outset, let us be clear. Each study will have its particular nuances. In some cases, new variables will have to be added to capture the relevant history and dynamics. Conversely, some of the factors cited in this guide may not be relevant to a particular case. Additionally, there may not be much publicly available information (or information available in English or some other accessible language) discussing some of the factors discussed in this guide, and time contraints may not permit the consideration of each of the variables discussed in this guide. This is unavoidable, and the writer should not despair. It is simply part of the process of writing such a case study. What we seek to do here is to provide

4

you with a tool that helps you understand where to begin exploring your particular case and to suggest potential variables to consider during your analysis.

Research Questions and Methodology

In general, most of the ARIS products rely on the case study method. Case studies are rich, contextual histories that explore a number of variables through qualitative methods. By standardizing the approach, eventually there will be a sufficient body of ARIS work to allow scholars to compare and contrast cases, which will yield novel insights. These insights will inform and cohere into best practices, which will help prepare the force.

Additionally, because the work is standardized, it is well suited for the classroom. This allows the Special Forces soldier to better prepare to carry out the UW mission by systematically exploring a set of cases. When real-world experience is not readily available, the best the soldier can do is study history. History becomes his laboratory and means of exploration. This is one of the underlying purposes of the ARIS work: to better educate the force to carry out the UW mission when called upon to do so.

Some of the animating research questions of this work are:

- Why do states that are external to a conflict use indigenous forces to achieve their strategic objectives?
- Once states make the decision to use indigenous forces, what types of support are most effective?
- If there are multiple indigenous groups an outside state can sponsor, how do sponsors determine which group to support? Do they support multiple groups?
- What types of control and oversight methods are instituted by external sponsors? How much decision-making authority is maintained by the external sponsor, and how much is devolved to the resistance movement?

THE EXTERNAL SUPPORTER

As you develop your particular case study, this section provides an overarching view of the conflict. The goal is to understand why the external supporter provided assistance and to set the stage for a detailed examination of the activities used to support the resistance movement.

Description

Provide a broad overview of the external supporter. In later sections, you will develop specific aspects in great detail. Your goal here is to produce a general overview that prepares the reader for greater detail later.

One factor to consider is to what degree an external sponsor is motivated by ideology and founding myths to provide support to a resistance movement. In the case of Iran, preceding the overthrow of the Shah, a revolutionary generation of clerics reinterpreted the "quietist" tradition of Shia Islam and began to view canonical figures at the dawn of the religion as revolutionary figures that challenged illegitimate and oppressive rulers. This reinterpretation in turn motivated the Islamic revolution and led to an effort to export the revolution to neighboring countries, such as Lebanon, which possessed a large Shia community. The broader point is that ideology motivated a foreign policy orientation, which in turn led the group to sponsor the establishment of Hizbollah

Additionally, one should keep in mind that Section III, the section on historical context, will primarily deal with historical factors that affect the country in which the insurgency took place rather than the history of the external sponsor. Hence, one can use Section II as an opportunity to describe any historical factors within the sponsor itself that help explain the provision of support to a resistance movement at a particular point in time. Thus, to return to the example of Iran, one cannot fully understand the motivation to sponsor Hizbollah without first understanding the conditions that led to the resistance of the Shah since Iran's Islamic revolutionaries were motivated by a set of grievances and ideas that also resonated with the founders of Hizbollah.

Strategic Goals

Next, the writer should describe the strategic context of the case study and in a general manner detail the international environment, the regional geopolitical landscape, and the strategic goals of the external supporter as they relate to the conflict under consideration. This section should include an overview of the external supporter's foreign policy, its strategic interests in region, past actions taken, and its relationships with various organizations, actors, and nation-states that are relevant to the study.

Interest in the Resistance Movement

Describe the external supporter's specific interests in the state where the conflict is taking place. Why is it interested in a particular resistance movement? Describe how working with the resistance movement helps it advance its strategic objectives. Examine the intentions and motivations of the external supporter in great detail, to include the specific goals it wants to reach through enabling a resistance movement. Be aware that these goals may be focused outside the nation-state where the resistance movement or conflict is raging, since support of a conflict in one country may enable the attainment of regional or strategic goals in another. For instance, a goal may be to create a buffer for an ally in a neighboring country or to create a capability that can be employed in a neighboring country or to undermine the interests and foreign policy of a neighboring state or regional or global adversary.

Resistance Movement Selection

Explain why an external actor provided support for a resistance movement or movements. This section should look at the alignment of the goals of the external actor and resistance movement or, in some cases, movements. Also examine other key aspects that may have led to external support for the conflict such as shared ideology, religion, or culture. This section should highlight the decision-making criteria used by the external supporter, which should include environmental factors, the viability of the resistance movement, any ideological affinities with the selected

7

resistance movement(s), and other courses of action that were available to the external supporter at the time.

HISTORICAL CONTEXT

The Historical Context section of the case study is important because it provides the reader with background information on elements within the country in which the resistance took place that are relevant to the development, objectives, and vulnerabilities of a resistance movement. The subtopics discussed within this section—physical environment, socioeconomic conditions, and government and politics—are addressed because these factors can have a significant influence on how and why a resistance movement develops, and additionally, these variables shape the goals, organization, strategies, and tactics of a resistance movement. The list is not exhaustive and new elements may be introduced based on the facts of the case being studied.

The variables within the Historical Context section are interrelated and affect the dynamics between the resistance group and other actors. The association between two or more variables may factor significantly into the analysis in subsequent sections of the case study. For example, the physical environment often impacts state capacity and development, which can affect popular support for resistance movements and the ability of the state to respond to challenges. From an external supporter's perspective, this type of information is important because it helps to identify the vulnerabilities of a resistance movement, which impacts the supporter's motivations and methods of support. Therefore, details on the physical environment, socioeconomic conditions, and the government's ability to provide basic services, among others, are necessary to include early in the study. A detailed understanding of these variables is the foundation the author will need to identify and expand upon associations that may explain the behavior of either the resistance movement or the external supporter. For the reader, they provide the context necessary to understand this analysis in later sections of the study.

Physical Environment

A description of the physical environment where a resistance movement or insurgency begins and develops is important because, at a basic level, resistance movements strategically use geographic features to increase survivability and to launch and resource operations. Insurgent groups such as the Afghan Taliban have benefitted from Afghanistan's mountainous terrain, which makes pursuit and surveillance by countervailing forces difficult. However, geography can be a double-edged sword. The physical environment may also present a vulnerability an outside actor can exploit. For instance, if the geography is rural, mountainous terrain with little arable land, providing support in the form of physical supplies may be difficult because the area is not easily accessible. At the same time, these factors may also impact the forces of a resisted government, so overcoming that difficultly and providing materials and basic necessities that are otherwise scarce may be one of the most effective ways for an outside actor to influence a resistance movement and the local population.

Other geographic features, such as location and distance, have also been found to impact conflict patterns and processes. Generally, regions farther from the state capital and closer to international borders are at higher risk for conflict. The mechanisms underlying the relationship between geographic location and insurgent activity are not clear-cut. In 1962, Kenneth Boulding developed a theory, the loss-of-strength gradient (LSG), according to which a state's ability to project power declines the farther it gets from its seat of power, the national capital. As a result, the state has less capacity and presence in peripheral regions, providing an opportunity for rebels to emerge and grow. [2] The LSG captures the importance of poor state presence, especially poor military presence, in peripheral regions, but a lack of state capacity also means that the regions suffer from poor social provisions, such as education and health care, which are well-known contributors to insurgent activity. Proximity to international borders offers support in neighboring countries and sanctuaries shielded by sovereign borders.

The external supporter must therefore determine how the physical environment plays a role in the resistance movement's development and trajectory and also how it creates vulnerabilities that may make external support effective. In the Physical Environment section, consider inclusion of the subsections listed below.

Terrain

This description should include the basic geographic features of the area or areas relevant to the resistance movement, which may extend beyond state or territorial borders. It should include features of the urban geography as appropriate. In the past, resistance movements either began, or solidified, their presence in the countryside on the peripherals of state power; however, many groups now develop within urban centers. The author should consider how aspects of the terrain factor into other sections and include the level of detail appropriate for that analysis. Rough terrain may be far from the urban areas and lack the basic infrastructure, such as roads, necessary for a robust state presence. As a result, a government may have little, if any, authority over residents there, which can create opportunities for insurgencies to emerge and flourish. Poor state capacity in these regions can also contribute to socioeconomic conditions that fuel rebellion, primarily the lack of infrastructure; the lack of basic social services such as health care and education; and poor economic opportunities. If the area is largely rural, geography may be a feature that creates grievances within the population, such as disputes over grazing rights or water rights, or lack of access to political processes that are carried out in distant urban locales. The description of the terrain should be adequate to later draw out these relationships to the extent the individual case study requires.

Climate

A description of an area's climate is important for reasons similar to those related to terrain. The area's general type of weather and seasonal changes may have an impact on a resistance group and when during the year it carries out certain activities. Climate may also relate to vulnerabilities within the population that are key to the external supporter. For instance, lack of access to water resources may present the outside actor with a form of support that would be uniquely valuable in gaining a foothold within the resistance movement or local population. Or, disputes over water rights within agricultural communities in dry climates may be an opportunity for an external actor to capitalize on a splintering population and provoke further disorder if doing so aligns with its motivations.

Transportation and Communications Infrastructure

The area's transportation and communications infrastructure are important variables for the author to describe because they are markers of how easily government personnel, security forces, other groups, and the external supporter can access the resistance movement. Primary passages or transportation points may be of strategic importance for infiltration. Who has primary control over the communication and transportation infrastructure is also a key point to note. If principal transportation and communication nodes are controlled by the government, the ability of the resistance group or an outside actor to make use of this infrastructure may be very limited. Beyond that, it may be an indicator of the government's level of control over the geographic area generally. If the government or government-controlled companies have a monopoly on the use of computer networks, broadcast towers, and supporting operations centers, then the state can control the information and messages disseminated to the public. It can use this infrastructure to influence messaging in the media about the resistance group or others. Infiltration of the infrastructure for use by the resistance movement or an outside actor becomes more difficult, which is why the author should include this description. The infrastructure may benefit or hinder the support and should be described with this focus in mind.

Geographic Scope of the Resistance Movement

Within the physical environment description, the author should describe the geographic scope of the resistance movement and its main areas of operation. This description should provide an outline of the region and neighboring states or territories. For instance, some resistance groups may have training camps in one country, enclaves in another, and be bordered by a supporting or unfriendly state. The external supporter needs to understand the regional influences on the resistance group. A neighboring state may provide sanctuary to the group in the form of operational bases or even protection from extradition. If these considerations are relevant to the given case study, the regional description will provide the necessary context for later analysis.

Socioeconomic Conditions

Although low economic development has been demonstrated to be a risk factor for rebellion, it is considered an indirect, contributing factor. Many societies have low levels of economic and human development, but the vast majority of individuals never take up arms to redress these grievances. However, poverty can help set the stage for political violence. Poor economic development means that many people in a country lack basic services, which also means the state has less ability to effectively counter challenges to its authority. The author should present enough information on the country's social and economic characteristics to explain to readers how the intersection of these conditions contributes to the political conflict and how socioeconomic conditions factor into an external actor's reasons for supporting a resistance movement. To provide the reader with adequate background, the author should describe the social characteristics and economic indicators separately, because either can be a source of division or conflict, and then explain how these factors overlap to create exploitable vulnerabilities within the population.

Social Characteristic/Demographics

The author should explain the general population demographics, including the size, growth, distribution of the population and whether there are significant numbers of refugees or internally displaced persons. The description should identify major groups that comprise the social structure and their identities, such as race, ethnicity, and religion. Common languages or dialects are also important. These cultural dimensions of the social structure affect attitudes and behaviors within a given population, such as the development of formal or informal relationships. The relationships between groups within the social structure may generate cleavages or become ties that cut across two or more aspects of identity. For example, a religious tie may unite groups despite ethnic differences.

Cultural dimensions may also be indicators of the level of modernization within the society, as reflected by the expectations of behavior or the status of certain groups within society. For instance, behavioral norms may apply exclusively to women or elevated social position may be based solely on tribal or ethnic background. It is important to note whether the resistance movement is composed largely of one social group or several and whether its composition differs from or aligns with that of the larger population. What is important to recognize is that social cleavages and demographic pressures can create social strife and conflict, which can be exploited by resistance movements.

Description of the Economy

The level of economic development is considered a contributing factor in conflict. The author should explain basic economic conditions and factors that contribute toward grievances and generate vulnerabilities; these conditions and factors may include the monetary system, major industries (licit and illicit), the unemployment rate and wages (particularly across different social groups), government deficit levels, gross domestic product, major trading partners, inflation rates, and per capita income, among others. In addition to these basic elements, the description should identify characteristics of the country's economy that are relevant to the creation of discontent in the population and that may serve as indicators of government effectiveness and control. For instance, poverty, unemployment, or poor health care may lead to social unrest, such as protests and rioting, and demonstrate a low level of government effectiveness caused by its inability to provide basic services. The author should include a discussion of historical or current economic crises that make the government particularly vulnerable to opposition. An external supporter may take advantage of a country's economy in order to provide support to the resistance movement.

While the level of economic development alone may not directly lead to the emergence of a sustained resistance, it can create opportunities for individuals to coalesce around a shared cause, eventually leading to a more organized movement. Additionally, "political entrepreneurs" looking to mobilize local populations can use these grievances as opportunities to make persuasive calls for collective action. Young men might feel they have few options in an economically depressed society, making participation in a resistance movement more attractive. Poverty also tends to promote criminal activity, desensitizing local populations to crime and building the foundations for a criminal class and illegal markets that often support insurgencies. Economic disparities can be exploited by an outside actor to its advantage. It is possible there is a historical tie the external supporter has to the country or its level of economic development, and any such ties should be included in the description because they may give the external actor particular influence over portions of the population that can be used to fuel a resistance movement.

Nexus of Socioeconomic Conditions and Creation of Vulnerabilities

The relationship between social characteristics and economic conditions is important to external support because often these combine to create vulnerabilities within a state and its population that a resistance movement or outside actor can exploit. In his book, *Why Men Rebel*, Ted Gurr developed a theory to explain the relationship between poor economic development and political violence.[3] He called it "relative deprivation," a theory that discontent arises when there is a mismatch between a group's expectation of what it should have and its actual economic reality. Relative deprivation is most pronounced among groups or people that have experienced recent surges in fortune (which in turn raise expectations) but find themselves still lagging behind their peer groups.[a] Political discontent among youth in several Middle Eastern countries, such as Egypt and Tunisia, fits the pattern of relative deprivation well.

---

[a] The truly impoverished, by contrast, often have little leisure or incentive to pursue political change.

Another important socioeconomic condition to consider is the intersection of ethnic identity and social and economic discrimination. When ethnic diversity is married with economic and political exclusion, the risk for political violence does increase.[4] Conflict is especially likely when a state government is ruled by one ethnic group at the exclusion of other groups. This is the case for several states that have recently undergone conflict, such as Sri Lanka, where the Sinhalese controlled most government institutions at the expense of the Tamils. Taking another example, the purposeful exclusion of Catholics from the Protestant-dominated government helped fuel both violent and nonviolent opposition to the state government in Northern Ireland. In the latter case, and in many other cases, political exclusion is accompanied by economic and social discrimination that translates into lower employment, education, and health care standards for excluded communities.[b]

An outside actor may use this vulnerability to help sell its support to the identified resistance group and to maintain a level of control over the supported group. Social and economic crises, such as ethnic violence, famine, economic sanctions, or a currency collapse, may provide windows of opportunity for resistance movements and their supporters. For example, if a particular ethnic group is losing ground in an internal conflict with other ethnic groups, external support may be readily accepted and even sought to prevent more loss or to shift from defensive to offensive actions. As you develop this section, describe and explain the vulnerabilities that exist both in the society and the economy. You should also examine how socioeconomic pressures and crises impact the government and the government's control and legitimacy among the

---

[b] One reliable source for identifying ethnic groups facing discrimination by their governments is the Minorities at Risk project. The project currently tracks approximately 280 minority groups at risk for discrimination, and each group is scored according to the levels of possible discrimination. In a testament to how influential this data set has become in studies of political violence and ethnic discrimination, researchers involved with the project have reported that ethnic groups from around the world have actively lobbied for inclusion in the list. It is important to note, however, that not all political conflict is driven by issues of ethnic identity. For instance, much of the political conflict in Latin America has been driven more by class-related, not ethnic, issues.

population. Crises create vulnerabilities in the government as well as in the population.

Government and Politics

The government and politics of a country can profoundly affect the emergence and trajectory of conflict there. This section should detail the government structure and political institutions within the country within which the resistance movement operates. The author should bear in mind that if the resistance movement has significant operations in several countries, it may be necessary to expand this section to include a description of the governments in each country or area. Unique characteristics may exist within each state's governmental system that are worthy of discussion, particularly if they create an appealing environment for the resistance movement to take root in or expand in.

To determine which variables to include in the Government and Politics section, the author should consider what aspects of the system are vulnerable to infiltration by the resistance or an external supporter; the extent to which the government has control in the area and over the population; and what aspects of the system, such as policies or insufficient capacity to govern, create motivations and opportunities for a resistance movement to emerge. For instance, a government that is unable to meet its population's basic needs may create an opportunity for a resistance group to step in and attempt to fulfill the government function or at least exploit the population's dissatisfaction to gain support. Further, a government that does not have control over portions of its territory or population may be incapable of stopping a resistance movement and its supporters. The writer should explain the willingness and the ability of the government to respond to, adapt to, reform in response to, and defeat the resistance movement.

In addition to the vulnerabilities of the population, the rigidity and weakness of the government and the extent of government control are important factors to external supporters when choosing which resistance movement to support. Furthermore, when resistance movements or insurgencies increase their challenges to the government and increase their responsibilities to the population, their requirements for resources and external support increase commensurately. These shifts in the development and phases of the resistance movement or insurgency create opportunities for external supporters to increase the depth and breadth of their relationship with the resistance movement as well as the general population. The relationship between civilian leadership and the state's security forces may also be important to understanding the government's willingness and ability to respond to a challenge posed by a resistance movement and its supporters. The following list describes several variables an author should consider.

Current Political System

The author should include a description of the political system of the country in which the resistance movement primarily operates. The reader will need to understand how the political system operates in order to grasp why the resistance movement opposes it and what elements are of interest to an external supporter seeking to address political objectives through its backing of the resistance. The description may include a discussion of several characteristics of the political system, such as:

- The type of government and its structure, including how it divides functions among various branches or divisions and how the leadership is selected;
- The political authorities and how they function;
- Major government policies, particularly any that create disenfranchisement;
- Participants in the political process and whether the process is accessible to broad portions of the population;
- Distribution of power between different groups and how imbalances in this distribution contribute to friction within the system and dissatisfaction among the population;
- Any element of polarization within the political system that manifests as extreme divergence of ideas between parties or groups within the system;

17

- The capacity of the government to provide basic services, such as sanitation services, transportation infrastructure, and schools; and

- The willingness of the government to address grievances through reform.

These characteristics are important because they indicate whether a government is vulnerable to a resistance movement. These characteristics also describe the relationship between the government and the governed and point to areas where they are susceptible to infiltration, subversion, and manipulation.

One should note that most researchers agree that highly developed, economically secure democratic states are the least vulnerable to political conflict. That is not to say, however, that highly repressive, authoritarian regimes are the most vulnerable to political conflict. Secure democracies provide pressure valves for societal discontent through well-trod legal institutional channels. In the United States, for instance, citizens are able to vote leaders out of office, contribute to groups lobbying for their interests, and engage in civil resistance to voice their discontent. These avenues for opposition ensure that discontent is managed through peaceful, legal channels. If radicalized resistance movements do opt to use violent or illegal means to achieve their political objectives, they will have difficulty raising support. For the average citizen, the costs are simply too high and the expected payoff too low.

In highly repressive regimes, the situation is nearly a mirror opposite of the situation facing open democratic societies. Highly repressive regimes provide no legal channels for political opposition or dissent. In these authoritarian states, it is difficult for political dissenters to form an organized political opposition to the regime. These regimes usually have highly refined secret police and other intelligence-gathering capabilities. Before the Syrian civil war and the Arab Spring, for instance, the Assad regime kept dissent in check through its secret police, the Mukhabarat. The police had an extensive intelligence apparatus supplemented by ordinary civilians encouraged to tattle on family, friends, and colleagues. As a result, most Syrians were highly suspicious of voicing dissent against the Assad regime.[5] In such regimes, any attempts at opposition are usually

18

met with arbitrary arrests, interrogations, and detentions. Political opposition is usually stillborn, crushed by the overwhelming force of the state's security apparatus. For the average citizen in these repressive regimes, such as North Korea, the costs of resistance are simply too high.

However, in today's world, many states fall somewhere between these two extremes. Social scientists call these states, which combine democratic and authoritarian features, hybrid regimes, or anocracies. These states might, for instance, have nominally democratic elections but might rig or otherwise corrupt election results. As a result, the ruling party or political leaders never face serious challenges to their authority. Some states, such as China, combine this restricted political freedom with greater economic freedom.

Researchers find that political conflict is more likely to arise in these anocracies than in truly democratic or repressive states.[6] This finding is referred to as the "inverted U-curve" because the distribution of political conflict on the authoritarian–democratic scale falls in the middle. These states typically allow just enough political and civil liberties that political opposition is able to form. The inherent contradictions in these states, which claim to be democratic but engage in activities that do not support these claims, also fuel societal grievances. When the political opposition mounts a challenge to the state, security forces often violently suppress the opposition, leading some resistance movements to adopt violence as a strategy to achieve their political objectives.[7]

Legitimacy

Writers may also want to consider the degree to which a populace considers existing political arrangements legitimate. Legitimacy is a difficult concept to define and measure, as it is an entirely subjective perception held by groups or populations in a country. Yet, legitimacy remains a central theme in the narratives of many resistance movements. Governments are legitimate when authorities are widely perceived by society as not only having the right to rule and expect obedience but also as having the capacity to fulfill the basic functions of government. One of the fundamental functions of governments is to provide security and safety to its citizens. When people fear for their safety, especially from threats such as terrorism, the perceived legitimacy of the government erodes. Justice is also an important component of legitimacy.

States with security and judicial institutions plagued with corruption and inefficiencies often experience low levels of perceived legitimacy among the population. A failure to deliver equitable settlements to disputes among the population leaves opportunities for alternative judicial systems like the informal Sharia courts in Afghanistan and Somalia. These "workarounds" further erode the legitimacy of the central government.

People also expect the government, to varying extents, to mirror their fundamental or sacred values. Sometimes, as in the case of many Muslim-majority countries, people expect the government to uphold and reflect religious values. In other countries, closely held secular values, such as freedom of speech and assembly, might also be crucially important. When governments fail to meet this litmus test of ideological legitimacy, political opposition is more likely.[8]

External Support of the Government

It is important to describe any external support the government receives, either through development funds, trade alliances, materials, or other forms, because this support can be used to counter a resistance movement. Moreover, an external supporter will be interested in a country's outside relationships because any outside support it receives is essentially a counter to its own support. It may be a factor an outside actor considers when determining whether to support a resistance movement.

Narrative of Critical Political Events

The author should include a discussion of critical political events that shaped or transformed the political system because these events may explain the current structure or political dynamics or highlight historical vulnerabilities that can be recreated or used by an external supporter when developing its strategy and objectives. In this regard, writers should be mindful of any long-running political processes that contributed to the emergence of political conflict in the country. Studying only recent observations of events in Northern Ireland, for instance, does not give the fullest picture of how and why armed conflict emerged there. As the centuries-long struggle with Britain in Ireland, and later Northern Ireland, illustrates, the political processes that gave rise to endemic conflict can be generations in the making. In these instances, historical reconstruction of these processes is necessary to understand the evolution of resistance movements. [9]

Furthermore, writers may want to consider the extent to which current political problems are due to stunted state and nation building. The state is the predominant form of political organization in the world today. State building looks different from region to region. In Europe, the pressures of interstate wars helped political authorities concentrate rule under a central government. [10] In other regions, such as Latin America, conflict between parties within the state, alongside international rivalries (short of war), helped concentrate power in a central government. [11]

Many modern states are also built on former colonial territories once held by European powers. In the wake of both world wars, victorious Western nations carved many nation-states from the defunct Ottoman Empire and colonial territories. The result is a multitude of states with arbitrarily drawn borders that do not reflect any national sentiments among their citizens. This legacy has continued to plague these states, especially in sub-Saharan Africa. Regardless of the paths state building took, the resulting constellation and relationship between governing institutions and between state and society profoundly affected politics downstream.

Military and Police Characteristics

The military and police characteristics may reflect the type of government in place and the extent to which the government can be infiltrated. Beyond that, they may demonstrate the level of control the government has over the population and how easily a resistance can gain momentum. The author should describe the attitude, morale, size, structure, and budget of the military and police forces. It is also important to note whether law enforcement conducts policing or if this is a function assumed by the military police. The level of repression exerted by these forces on the population and any evidence of corruption are also important to include. The author should describe the intelligence activities of the government and which group is responsible for their oversight and execution. For example, if a state has a highly refined secret police or other intelligence gathering capabilities, it is very difficult for political dissenters to form an organized political opposition to the regime.

An external supporter will need to consider how these characteristics will affect their support and what methods of support will be most effective. A description of military and police characteristics should analyze whether the government truly has control over the internal security of the country or whether it only maintains control in certain areas, such as major cities. The reach of the military, police, and intelligence capabilities is also critical because an extensive network may present major difficulties for infiltration by an outside actor, but a limited network in more rural areas presents an opportunity for access into the country. The external supporter will need to determine how it can gain access to the country, and the government's military or police presence is one consideration. The author should refer back to the physical geography section as appropriate to highlight here how those features relate to government control and outside access.

The relationship between civilians and the military and police forces is relevant to this discussion because it may indicate the popular support base of the government and also the degree to which the population trusts or mistrusts these institutions. It is possible the military and police forces target certain demographic groups in order to coerce or manipulate them to gain or maintain control in an area. This idea relates to corruption and should be tied to any discussion related to corruption.

The Level and Extent of Government Control

The level and extent of government control are important factors in determining whether a resistance movement can grow and be successful. With respect to external support for a movement, government control will affect the decision calculus of outside supporters. Although government control is not a sole determinant of external support, it does matter. To what level it matters is part of what must be explored.

Other areas of exploration could detail the level and extent of government control as it relates to its population and territory. Describe the physical and geographic limitations of government control, if there are any. Additionally, it may be useful to provide an accounting of the following: the portion of the population that is loyal to the government for both economic and other reasons; major government facilities and bases; and border control points and their impact on security. These are some indicators of the sitting government's control and effectiveness.

THE RESISTANCE MOVEMENT

This section looks at the capabilities and vulnerabilities of the groups participating in the resistance movement. Additionally, one will want to describe the drivers of the conflict and how they are exploited by the resistance movement and the external supporter. Describe the events that brought the external supporter and resistance movement together, and explore the ways the external supporter may influence or leverage the resistance movement. Detail the attributes of the resistance movement and how they shape the actions of the external supporter.

Nature of the Resistance Movement or Insurgency

The writer should describe the nature of the resistance movement or insurgency. This description should include the origins of the resistance movement or insurgency, the strategic and overarching goals and objectives of the resistance movement or insurgency, and the type of resistance movement or insurgency. For example, is it a secessionist insurgency or a resistance movement to expel a foreign occupying force? Does the resistance movement or insurgency wish to institute a Marxist political system or a caliphate? This section should describe the ideology, philosophy, or political theory of the resistance movement. The writer should also explain whether these goals align with those of the external supporter and how and why they align. Further, describe any involvement the external supporter had in the origins of the resistance movement or insurgency, such as creating or exacerbating socioeconomic conditions that gave rise to resistance or insurgency in the country.

In discussing a group's ideology, one should note that some groups adopt highly systematized ideologies, whereas others operate under more loosely defined goals. In either case, though, ideologies serve both instrumental and normative functions. The former is the most common approach to understanding the role of ideologies in a resistance movement. In this context, the term *instrumental* describes how ideologies serve a useful, strategic purpose. Some important instrumental tasks writers should consider are mobilization, indoctrination, and support; however, this list is by no means exhaustive.

Secondly, ideologies serve an important normative function. This function is less well understood than the instrumental one discussed above. In this context, the term *normative* means that an ideology can constrain or direct behavior outside of strategic purposes through the sheer "power of ideas." In this section, writers should analyze to what extent a group's ideology constrains or encourages different behaviors. A highly systematized ideology, such as the Marxist ideologies many insurgencies adopted in the developing world, can come prepackaged with prescriptions for different institutions and future political regimes. [12] A nationalist ideology usually also prescribes some form of secession or autonomy from a central government.

Acceptable and unacceptable forms of violence can also be directed by a group's ideology. One barrier that groups must often overcome is a communal abhorrence of violence. [13] A member of the April 19th Movement (M-19) in Colombia described how the group's Marxist-derived ideology helped her conceptualize violence as a moral imperative necessary to achieve social justice. Likewise, groups espousing Islamic ideology must reinterpret sacred scripture to support otherwise unacceptable practices such as suicide bombings and the killing of innocent civilians.

Strategies and Supporting Narratives

Here, the writer should describe the strategies used by the resistance movement. For example, does the resistance movement use a protracted popular warfare strategy or a military-focused strategy? The description should include the use of religion, ideology, tribalism, or nationalism in the movement's strategy and narrative. For instance, a resistance movement or insurgency that has co-opted a religion and has embedded itself in the religious infrastructure in a particular area may draw significant advantage in protection, legitimacy, and obedience from this relationship with religion, but this relationship may also limit the inclusion of other segments of the population and limit potential relationships with external supporters.

Writers should also be aware that important political happenings, such as revolutions, protests, or other events, are lodged in a society's collective memory as names, symbols, or other images. Resistance movements make use of these historical memories as signals communicating intent, as threats lobbed at political authorities, or as models on which to build future activities. In the case of the Solidarity movement in Poland that unseated the ruling Communist regime in 1989, the trade union frequently referenced a series of dates, including 1956, 1968, 1970, and 1976. Earlier in Polish history, different classes of Poles—students, workers, and intellectuals—had independently launched protest movements against the regime in each of these years. Authorities ruthlessly suppressed each protest in turn, sometimes resorting to armed violence against unarmed protestors.[14] Simply by referencing the dates of these events, the Solidarity movement communicated a rich narrative to its audience. The dates were replete with remembrances of past injustices and the failure to work together and provided inspiration to take up where previous protestors had failed. Thus, explaining the Solidarity movement requires a historical reconstruction of not only what happened in those failed protests but also how the Solidarity leadership used the collective memory of those dates to mobilize and inspire the Polish public.

This section should also look at how a resistance movement exploits, and even creates, vulnerabilities in the population, weakening government control in the area. Explore how factors in the human and physical environment as well as the political environment impact the movement's strategy. Include the strategic use of narratives and the inclusion of cultural or symbolic aspects of the society or geography in those narratives.[c] The writer should explain why the strategies are important to the external supporter, if they are, and note any influence by the external supporter in shaping the movement's strategy.

Structure and Dynamics of the Resistance Movement

Leadership

This section should focus on the key leaders of the resistance movement, their strengths and weaknesses, and the role of leadership within the organization. The writer should include the style of the leaders, their roles, activities, associates, personal background, personal beliefs, motivations, ideology, education, training, temperament, position within the organization, and popularity, to name a few. Describe the impact that key leaders have had on the resistance movement. Explain the leadership structures that shape the resistance movement. Additionally, describe how leaders are selected and how they exercise their power. The writer should look for ways external supporters have influenced the selection of leaders and the decisions of leaders, eliminated leaders, and pressured leaders to serve their strategic goals.

Organizational Structure

Describe how organized the resistance movement is, and the typology and characteristics of the organization, whether it is centralized or decentralized, its size, and its organizational capacity to perform all of the required functions needed to be a viable resistance movement or insurgency. It is important to understand its organizational capacity: is it only a small opposition group or is it capable of governing the country if successful? The writer should include how the organization provides the supplies, transportation, communications, medical, propaganda,

---

[c] One may want to employ the framework provided by social movement theory, with its emphasis on diagnostic, prognostic, and motivational frames, to analyze the narratives of groups that employ UW tactics. For more information on social movement theory, see Robert D. Benford and David A. Snow, "Framing Processes and Social Movements: An Overview and Assessment," *Annual Review of Sociology* 26 (2000): 611–639.

intelligence, counterintelligence, recruiting, and finance needed to support the resistance movement.

Also, all political and military structures should be described along with their overt, covert, and clandestine substructures and capabilities, i.e., describe the resistance movement's underground, auxiliary, and guerrillas, as well as any shadow or parallel governments. Describe how they operate in rural areas as well as urban areas, and describe how the human and physical environment as well as the political environment shapes their organizational design, structure, and reach. This section should include all of the physical as well as virtual infrastructure developed and used by the resistance movement. The writer should describe the organization in as much detail as possible to understand the its vulnerabilities and any weakness in its capacity that can be either bolstered or exploited by external supporters.

Additionally, one should be aware that most command and control structures involve numerous trade-offs, especially among control, efficiency, and secrecy. A hierarchical structure ensures consistency and control across multiple functions, but it also produces bottlenecks that can slow response time. Flat or networked organizations might sacrifice some control but gain efficiencies in the absence of the layers of authority that make up hierarchical systems. In organizations for which mistakes are especially costly, the consistency offered by a hierarchical structure might trump the speed of response and the secrecy afforded in other structures. When the perception of legitimacy among a population is of particular concern, or during negotiations with authorities, needless civilian deaths or radical splinters from the group might prove intolerable for the organization. A group's ability to adapt its organizational structure can be a determinant of its survival and success.

Lastly, the writer should not only examine the organizational structure a resistance movement's leadership adopts but should also explore the ramifications of those decisions. At a most basic level, the organizational structure of a group can and does impact the operations carried out by the group. Structures shape incentives for different behaviors. Recent research has shown that those groups with more specialized units dedicated to violence commit more lethal attacks and sustain violence over a longer period of time. For example, the Basque group Euskadi Ta Askatasuna (Basque Homeland and Freedom, or ETA) was more violent during periods when its units were more specialized.[15] Likewise, structures can facilitate or constrain different objectives.

Command, Control, Communications, and Computers

Explain how the leaders of the organization command, communicate, and control the various parts of the organization. This should include direct and indirect communications, clandestine methods of communications, and the decision-making processes guiding the group's actions. Also describe the communication and information technology employed by the resistance movement to conduct its operations. Note how the external supporter provides capabilities in these areas, potentially leading to control over the organization or parts of the organization.

Geographic Extent of the Resistance Movement or Insurgency

Here the writer should define the geographic disposition of the resistance movement, making sure to differentiate between territory that is permissive, semipermissive, and nonpermissive for the movement. Show where the resistance movement has control over territory and population, where there is contested control, and where the government has control. Describe the extent of operations, locations of political and military actions and major events, and the locations of bases, training camps, safe houses, and sector and area commands. This section should focus on the physical aspects of the organization, and it should include the physical locations of external support sources and infiltration routes, both physical and virtual. It is important to get a sense of how the resistance movement has adapted to fit the physical geography of the country and region and any significant challenges posed by it. Also, it is important to get a sense of the proximity of external support to the resistance movement and the ease with which it can be provided. Lastly, ensure that areas of sanctuary are geographically depicted, including sanctuary in neighboring countries, refugee camps, and within the territory controlled by external supporters.

Resources and External Support

In this section, try to capture and describe the resources of the resistance movement and how the movement uses or employs the resources. For instance, what percentage of food supplies are used to support guerilla forces or used to build popular support? This section should include all of the funding, supplies, weapons and ammunition, and other materials the resistance movement has at its disposal. When possible, the actual quantities of resources and funding should be described. The writer should cover the sources of these materials, the distribution methods and supply lines, and stores or facilities. Additionally, describe and highlight the types of resources the resistance movement or insurgency has internally and which are provided by external supporters, both material and nonmaterial, such as technical expertise. Taking this one step further, the writer should cover all of the resources received from external sources, their providers, the location of those resources, the motivations for providing the resources, their methods of delivery, and which are the most critical for the survival and growth of the resistance movement or insurgency.

Political Activities

This section should include all of the nonviolent activities of the resistance movement. Suggested topics include all activities focused on the resistance movement's popular support base, the government's popular support base, and its rival's poplar support base. Activities may include subversion, agitation, passive resistance, demonstrations, strikes, boycotts, infiltration of government institutions, fundraising, cadre development and growth, and organization building. This should explain the operational strategies and goals of the organization and the tactics and techniques used to mobilize its popular support base and erode the government's popular support base and institutions. This will include use of mass media, social media, and propaganda and information activities. This may also include the provision of essential services, social and economic relief, and governance to segments of the population.

The writer should highlight the organization's ability and capacity to conduct political activities across the full range of possible activities. For instance, can the resistance movement or insurgency create and direct a nation-wide strike or only local strikes? Can the resistance movement or insurgency infiltrate and subvert the most important political, economic, and security institutions? The writer should also examine the organization's use and manipulation of spontaneous mass popular activities such as mobs and riots, noting which political activities are conducted to gain external support. Examine this in detail as well as the coordination of political activities between the resistance movement and the external supporter and the conduct of political activities at the request or demand of external supporters in exchange for support.

Methods of Warfare

This section should include all of the violent activities of the resistance movement. It will include all of the violent and forceful activities focused on the resistance movement's popular support base, the government's popular support base, and the rival movement's poplar support base, as well as other aspects of society, the economy, and the institutions of the government. This section should include all of the activities of the resistance movement's guerrillas, terrorists, paramilitary, militia, regular or conventional forces, and criminal wings. Activities may include targeted political assassinations, raids, ambushes, sabotage, improvised explosive devise use, suicide bombing, attacks against civilian populations, attacks against government civilian institutions, attacks against government police or security organizations, and attacks against the external supporters of the government. Terrorism and violent criminal activity such as kidnapping, extortion, coercion, and the use of violent threats to achieve organizational political and military goals should be included in this section.

This section should explain the operational strategies and goals of the organization as it concerns the use of violence to achieve its objectives. For example, some activities will be designed to demonstrate weakness in the government in the eyes of the population, some will be designed to provoke an overreaction by government forces, and some will be designed to diminish the capabilities of the government's forces. When possible, show how these operational strategies and goals nest within or contradict the organization's overall strategy and goals. Further, the writer should describe the military tactics used by the organization's forces and the weapons, munitions, and equipment used. The writer should also describe how the resistance movement or insurgency uses its forces to provide security and protection to segments of the population. This section should also describe the activities of each of the components of the resistance movement that uses violence, such as the guerrillas, auxiliary, and action cells.

The writer should highlight the organization's ability and capacity to conduct violent activities across the full range of possible activities. For instance, can the resistance movement or insurgency assassinate senior leaders in government or only local leaders? Can the resistance movement or insurgency sabotage or destroy national-level infrastructure in such a way that it disrupts the entire economy and large portions of the population, or can it only create localized effects with its actions? The writer should also examine the organization's ability to control territory by force, the capability and capacity for traditional warfare, and the ability to mobilize large segments of the population to participate in violent action against the government or rival opposition groups. As recommended in the Political Warfare section,the writer should also examine the organization's use and manipulation of spontaneous mass popular activities such as mobs and riots as tools to achieve military goals when analyzing methods of warfare. Since some violent activities will be conducted to gain external support, the writer should examine these in detail as well as the coordination of violent activities between the resistance movement and the external supporter and the use of violence at the request or demand of external supporters in exchange for support.

Popular Support for the Resistance Movement or Insurgency

Explore the relationships between the resistance movement and various segments of the population that support it. The writer should describe the social composition and level of popular support for the resistance movement or insurgency. This should include why the population accepts the organization, the limits of this acceptance, and the vulnerabilities within these relationships. For instance, does the resistance movement provide essential services to the population, and if the government provided the services, would the population then support it? The socioeconomic conditions and grievances exploited by the resistance movement or insurgency should be explained in detail here, to include the techniques used to gain popular support, how the resistance movement or insurgency gains and maintains legitimacy, and the willingness and tolerance of the population for violence against the government and in society in general.

The writer should include both passive and active support, techniques for mobilizing segments of the population for political or violent action, and the impact of the various political and violent activities of the resistance movement on the different segments of the population. Describe how the organization recruits and maintains its numbers from its support base. Some activities will negatively impact the organization's popular support but may be necessary for attaining other goals such as acquiring resources, international recognition, or external support. Another aspect to examine is the relationship between the external supporter and the population. External supporters can form a relationship directly with the population and use these ties as leverage to influence the resistance movement or to take unilateral action.

History of the Resistance Movement or Insurgency

Capture the history and evolution of the resistance movement. Include the major events of the conflict, battles, and changes throughout the struggle, and how, if at all, it is connected to earlier conflicts or other conflicts in the region. Address the phases of insurgency, how the organization shifted between phases, and how the organization adapted its strategies and goals as it passed from one phase to another. This section should also address the changes the organization made to its tactics and techniques and the impact of these changes on the course of the resistance movement. Also, describe how the resistance movement or insurgency overcame any significant challenges it faced in terms of leadership, resources, organization, and strategies and goals. This section should include how the government responded to the resistance movement and describe any actions taken by the government to eliminate external support to the resistance movement.

EXTERNAL SUPPORT

Explain and describe the nature of the relationship between the resistance movement and a specific external supporter. Develop the history of the relationship, to include any past interaction with the population prior to the development of the resistance movement. Elaborate on the total amount of external support provided and the conditions attached to the provision of support. Include the points at which the resistance movement sought and attained external support and the significance of the external support to the outcome of the conflict and the development of the resistance movement. Describe the results of the provision of external support for both the resistance movement and the external supporter. Develop the history of the relationship and the critical junctures in the relationship. This section should address all types of actors that the external supporter used, both witting and unwitting, and all of the forms of support.

Keep in mind that there may be various agencies within an external supporter that are involved in the relationship with the sponsored group(s). For instance, ministries of defense and foreign affairs, various intelligence agencies, different actors within the armed forces, and the cabinet may be involved in managing the relationship. One should therefore not assume that the external sponsor is a unitary actor, since each of the agencies involved in the relationship may have their own agendas and interests, and thus their actions may not be optimally coordinated. Additionally, if the external sponsor is characterized by factionalism among its key political elites, the factions may differ in terms of their preferred policy toward the resistance movement. A resistance movement may therefore develop a better relationship with one faction over another.

Lastly, when considering the plethora of potential actors involved in external support, one will want to be sensitive regarding which actor(s) actually possess decision-making authority regarding strategic and tactical issues affecting the resistance movement. Political scientists sometimes use the term "veto players" to indicate that a particular actor has decision-making authority over a specific issue. In the current context, one will want to note where the locus of power resides within the external actor regarding strategic and tactical decisions affecting the sponsored group as well as implementing policies that have been agreed to.

Type of Actor Providing Support

There are many types of actors that provide external support to a resistance movement. Foreign nation-states provide a great deal of external support as do diasporas, refugees, and other nonstate actors such as religious organizations, nongovernmental organizations, affluent individuals, other resistance movements, and criminal organizations. In this section, examine the external actor specific to the case study, ensuring to differentiate between other external actors. For instance, a state may provide a shipment of weapons and munitions directly to a resistance movement. A state may provide financial assistance or distribute cash directly to the resistance movement's leadership. A state may also use a religious organization to provide financial assistance by establishing a new religious organization specifically for the resistance movement. Be aware that a state may use a variety of actors to provide support. This can create a false sense of international support for a resistance movement, effectively masking the full extent of the state actor's support. Further, a state can use a number of external actors unwittingly or encourage more support from other external actors. Although it may be difficult to determine all of the hidden drivers of external support, try to discover and detail as many as possible.

Motivations for External Support

Develop and explore the various motivations of the external supporter. Include the motivations of the various external actors, large and small, that are part of the support network. Keep in mind that various third parties may play an important intermediating role between the external sponsor and the resistance movement. For instance, Syria plays a crucial role in the relationship between Iran and Hizbollah, especially since all Iranian aid had to first transit through Syria before reaching Hizbollah (traditionally Syria did not provide overflight rights for Iranian planes to directly fly to Lebanon). Taking another example, during the Cold War, Pakistan played an important role in intermediating American and Saudi aid to the Afghan resistance fighting Soviet forces following the Soviet invasion of Afghanistan. Hence, the interests and motivations of third parties may also need to be considered.

Motivations can be broadly grouped into four categories, strategic goals, benefits, capabilities, and affinities. Include things like a shared sense of religion, ethnicity, identity, or ideology, as well as shared political sentiments or a common enemy. These shared interests can form the basis of relationships between an external supporter and the resistance movement. State external supporters may also have strategic goals and regional interests that form the basis for their motivation to support a resistance movement, such as increasing regional influence, destabilizing other governments or rivals, changing the regime of another government, maintaining influence within a resistance movement, and maintaining their own internal security if the resistance movement or insurgency has a presence in the external supporter's territory.

Moreover, states may choose to support resistance movements when they are incapable of attaining their strategic goals through conventional warfare or diplomacy. A state may be unable to defeat a rival through conventional means and may choose an indirect form of warfare such as supporting a resistance movement within a rival's territory. Furthermore, the benefits of using a resistance movement as a surrogate or proxy to achieve strategic goals are appealing. Indirect warfare may reduce costs, may reduce political consequences, and may be easier to pursue than an overt invasion. States may avoid negative international or internal public pressure if the war is hidden and waged through a resistance movement.

This section should examine all of the motivations for providing support. Again, include the overarching strategic goals of the external supporter and link back to the section describing their strategic goals for the region. Understanding motivations can provide valuable insight into the reasons external supporters provide large amounts of resources to resistance movements for long periods of time and explain why external supporters often have difficulty extracting themselves from a conflict. Additionally, understanding the motivations and goals of the external supporter can provide insight into how one can defeat external supporters without having to defeat or contain all of the forms and methods of support.

Forms of Support

Describe the forms of support a resistance movement receives from an external supporter. Examples include state-provided materials such as weapons, munitions, and other supplies, as well as financial support, organizational assistance, expertise, training, or technological support, moral and political support, intelligence, propaganda materials, and sanctuary. External supporters may provide scarce resources to the resistance movement or insurgency that are critical to its success. External supporters can use the provision of resources and funding as major sources of leverage and influence over resistance movements or insurgencies. The resistance movement may become dependent on an external supporter for resources that are not available to it internally. This dependency can be used to control or influence the resistance movement or insurgency. The writer should examine in detail the dependencies created, exploited, and perpetuated by the external supporter.

Methods Used to Provide the Support

Explore and detail the various methods and techniques used to provide external support. Tie the actors and the section on forms of support together and explain in detail the mechanisms used to provide material and nonmaterial support. Describe the external supporter's organizations, institutions, special operations forces, front companies, and clandestine networks developed for providing external support. Also describe the legal authorities under which support is provided. Examine how the external supporter manipulates other nonstate actors to support the resistance movement. For instance, a state may raise awareness of factual or fictional human rights violations committed by the government opposed by the resistance movement. This may cause nongovernmental organizations, international organizations, diasporas, refugees, and other states to begin to provide various forms of external support to the resistance movement or insurgency. This additional support may be welcomed and it may provide new avenues of access for the external state supporter to exploit. Further, the additional support provides legitimacy for externally supporting a resistance movement. Include the ways in which the methods of providing support afford the external supporter leverage over the resistance movement.

CONCLUSION

In this section, address whether the external supporter attained any of its strategic goals or effectively pursued its interests by sponsoring a resistance movement, and summarize how. Consider whether the external supporter could have achieved its goals more effectively if it had chosen a different group to work with or if it had better managed the relationship with the chosen group (or if it could have more effectively achieved its goals without sponsoring any groups).

Also address points at which the external supporter's goals converged or diverged with the resistance movement's goals, examining any resulting strengthening or weakening of the relationship. Further, try to determine which actor drove the strategic agenda. Was it the external supporter or the resistance movement? For instance, a resistance movement or insurgency may adopt the global or regional goals of their external supporters or associates in order to gain needed resources. This adoption of a global agenda may in turn cause the resistance movement or insurgency to lose popular support of segments of the population that have a local agenda for change and do not desire to have their local agenda subordinated to a regional or global agenda set by a foreign entity. Additionally, a resistance movement may acquire new enemies if forced by the external sponsor to broaden its target set. However, external support can allow an unpopular resistance movement or insurgency to persist. Finally, gauge the effectiveness of the external support, i.e., explain which combinations of actors, forms, and methods of support were the most effective and why for the case under study.

APPENDIX: The UW Case Study Outline

I. **Introduction**
   a. Introduction
   b. Purpose
   c. Research Questions and Methodology

II. **The External Supporter**
   a. Description
   b. Strategic Goals
   c. Interest in the Resistance Movement
   d. Resistance Movement Selection

III. **The Historical Context**
   a. Physical Environment
       i. Terrain
       ii. Climate
       iii. Transportation and Communications Infrastructure
       iv. Geographic Scope of the Resistance Movement
   b. Socioeconomic Conditions
       i. Social Characteristic/Demographics
   c. Description of the Economy
   d. Nexus of Socioeconomic Conditions and Creation of Vulnerabilities
   e. Government and Politics
       i. Current Political System
       ii. Legitimacy
       iii. External Support of the Government
       iv. Narrative of Critical Political Events
       v. Military and Police Characteristics
       vi. The Level and Extent of Government Control

IV. **The Resistance Movement**
   a. Nature of the Resistance Movement
   b. Strategies and Supporting Narratives
   c. Structure and Dynamics of the Resistance Movement
       i. Leadership
       ii. Organizational Structure

           iii. Command Control Communications Computers (C4)

           iv. Geographic Extent of the Resistance Movement or Insurgency

           v. Resources and External Support

    d. Political Activities

    e. Methods of Warfare

    f. Popular Support for the Resistance Movement

    g. History of the Resistance Movement or Insurgency

V.   External Support

    a. Type of Actor Providing Support

    b. Motivations for External Support

    c. Forms of Support

    d. Methods Used to Provide the Support

VI.   Conclusion

NOTES

---

[1] Joint Publication 1-02, *DoD Dictionary of Military and Associated Terms* (US Joint Chiefs of Staff, 8 November 2010, as amended through 15 October 2013).

[2] Kenneth E. Boulding, *Conflict and Defense: A General Theory* (New York: Harper, 1962).

[3] Ted Robert Gurr, *Why Men Rebel* (Boulder, CO: Paradigm Publishers, 2011).

[4] Lars-Erik Cederman, Andreas Wimmer, and Brian Min, "Why do Ethnic Groups Rebel?: New Data and Analysis," *World Politics* 62, no. 1 (2010): 87–119.

[5] Lisa Wedeen, *Ambiguities of Domination: Politics, Rhetoric, and Symbols in Contemporary Syria* (Chicago: University of Chicago Press, 1999).

[6] Håvard Hegre, Tanja Ellingsen, Scott Gates, and Nils Petter Gleditsch, "Toward a Democratic Civil Peace?," *American Political Science Review* 95, no. 1 (2001): 33–48.

[7] Ibid.

[8] Nathan Bos, "Underlying Causes of Violence," in *Human Factors Considerations of Undergrounds in Insurgencies*, 2nd ed., ed. Nathan Bos (Ft. Bragg, NC: United States Army Special Operations Command, 2012), 19–20.

[9] Charles Tilly, "Why and How History Matters," in *The Oxford Handbook of Contextual Political Analysis*, eds. Robert E. Goodin and Charles Tilly (Oxford and New York: Oxford University Press, 2006), 420–421.

[10] Charles Tilly, "War Making and State Making as Organized Crime," in *Bringing the State Back In*, eds. Peter Evans, Dietrich Rueschemeyer, and Theda Skocpol (Cambridge: Cambridge University Press, 1985).

[11] Cameron G. Thies, "War, Rivalry, and State Building in Latin America," *American Journal of Political Science* 49, no. 3 (2005): 451–465.

[12] Stathis N. Kalyvas and Laia Balcells, "International System and Technologies of Rebellion: How the End of the Cold War Shaped Internal Conflict," *American Political Science Review* 104, no. 3 (2010): 415–429; and Francisco Gutierrez Sanın and Elisabeth Jean Wood, "Ideology in Civil War: Instrumental Adoption and Beyond," *Journal of Peace Research* 51, no. 2 (2014): 213–226.

[13] Robert Leonhard, *Undergrounds in Insurgent, Revolutionary, and Resistance Warfare*, 2nd ed. (Ft. Bragg, NC: United States Army Special Operations Command, 2013), 123–124.

[14] Chuck Crossett and Summer Newton, "Solidarity," in *Casebook on Insurgency and Revolutionary Warfare Volume II: 1962–2009*, ed. Chuck Crossett (Ft. Bragg, NC: United States Army Special Operations Command, 2012), 830–831.

[15] Bos, "Underlying Causes of Violence," 19–20.

41

www.ingramcontent.com/pod-product-compliance
Lightning Source LLC
Chambersburg PA
CBHW052118020426

42335CB00021B/2815